Baby Pet Animals

by RONNE PELTZMAN RANDALL
illustrated by SHEILA SMITH

LADYBIRD BOOKS, INC., Auburn, Maine 04210 U.S.A.

Printed in U.S.A.

Kittens have fun playing indoors...

...or out in the garden.

Baby pony has his own field to play in.

This puppy's favorite toy
is a soft old slipper.

These puppies have had a busy morning. Now they are thirsty!

The baby guinea pigs and their mother
have wiggly whiskers and noses.
What can they smell?

The baby rabbits
hear a bumblebee buzzing.

Someone has left a crisp carrot
for these baby rabbits to nibble.

The baby gerbils
are busy eating their lunch.

The baby parakeets have bells to ring,
and a ladder and swing.

Baby hamsters like
crunchy nuts and seeds
for supper.

Baby turtle likes to walk
on the fresh green grass.